CEYLON
STONE IMPLEMENTS

CEYLON
STONE IMPLEMENTS

JOHN POLE

GYAN

Published by

Gyan Publishing House
5, Ansari Road
Daryaganj, New Delhi-110002
Phone: 011-47034999, 9811692060
E-mail: books@gyanbooks.com

Distribution Network
gyanbooks.com
India, USA, Canada, UK, Australia, France

ISBN : 9788121232159 (PB)
First Published, Calcutta, 1913

2nd Impression 2020

Printed at: Gyan Press, Delhi.

Ceylon
Stone Implements
Author: John Pole

CEYLON
STONE IMPLEMENTS

By

JOHN POLE

SCARBOROUGH ESTATE

MASKELIYA

CEYLON

Printed by

THACKER, SPINK & CO

CALCUTTA

1913

PREFACE.

An attempt has been made, in this paper, to give a short account of what is at present known, of the Stone Age in Ceylon. It is particularly hoped that it may arouse an interest amongst those men who are in the best position to add to our knowledge of the subject. I refer to those Sinhalese Chieftains—the native gentry of the Island—who still find their pleasure in managing (often personally) their princely domains.

Little or nothing is known of the earliest types of implements that were fabricated in Ceylon. It has even been suggested that our quartz series of stones are not really prehistoric, but belong to an age that is recent—as compared with that generally attributed to similar implements in other lands. It is impossible, at present, to prove or disprove this theory, owing to the incompleteness of our knowledge. Even the fact that vestiges of a Stone Age in Ceylon exist, is a matter of comparatively recent recognition. It is now incumbent upon our native aristocracy to further this research on every possible opportunity. The construction of irrigation channels, the cutting of foundations or the removal of large bodies of soil for any purpose, the 'aswedumizing' of ancient lands and the various operations of bringing waste lands under cultivation, all offer opportunities that should not be neglected. Every stone of unusual or interesting form that may be unearthed during such operations should be secured and reserved for inspection.

Advantage should be taken of any disturbance of alluvial soil, such as is found in river beds and during the operation of gemming. Even the natural landslips that so often occur on the banks of rivers might prove of value in determining whether Palæolithic man of a vastly anterior age inhabited this Island and fabricated rough stone implements, fragments of which we believe that we possess, and of which it is anticipated that more perfect examples will be found beneath the surface soil.

In comparison with the abundance of artificially chipped quartz stones, worked chert flakes have been found but sparsely in the Maskeliya District, and most of these have been taken on the central chain of hills running through Bargany, Alton, Gouravilla and Maskeliya Estates. The distribution of chert itself is much more limited than that of quartz. Large blocks of this mineral are rare. Strangely enough, no chert implements were discovered on Clodagh Estate (in the Matale District), although chert occurred there in greater quantity than I have seen elsewhere. There are certainly no veins of chert in the Maskeliya District, but of the scattered pieces—averaging from four to five cubic inches—nearly everyone has been worked, or,—as one might say—"nibbled" at.* Puttalam provided one elongated flake with a single ridge, and a small double-ridged flake was found at Peradeniya. A few small flakes of a composite cherty conglomerate occurred at Mankulam. The Uva Province appears to have been the home of the workers in chert; for not only have their chippings been picked up on the patanas in the open, but large numbers have been dug out of the caves of the Eastern and Uva Provinces by the Doctors Sarasin. Mr. G. P. Cann obtained a single disc-shaped implement from a drain in Madulsima. Our Maskeliya examples are usually found on the ridges, after the ground has been broken up during the cultivation of the tea.

English authorities have proclaimed our quartz implements as of the Neolithic Age (some few, indeed, denying that they are even prehistoric) contrary to the opinion of the Drs. Sarasin that they are Palæolithic.

No notice of the Ceylon Stone Age would be complete without mention of the names of the two learned brothers, the

---

* Figured by the Drs Sarasin.

Doctors P. and Fritz Sarasin. For, had it not been for the repeated researches of these scientists upon the anthropology of the Veddahs, Ceylon archæologists would still be labouring under the mistaken idea that this Island had never passed through a Stone Age, and those persons who discredited our earlier collections would still hold the view that all vestiges of that age were wanting. As a matter of fact, the discoveries of the Sarasins in the Uva Province and elsewhere led us to believe that chips of white quartz and clear crystal are to be found all over Ceylon, and they have been found, from Mankulam in the north to Galle in the south, and Hambantota in the east. I myself am firmly convinced that they occur in equal abundance throughout the whole of the Central Province. In addition, stones typical of an earlier age (the cave and river-drift) and somewhat similar to the palæoliths from Madras, have been found in the Maskeliya district. Mr. Bruce Foote had already assured me that such would be found—sooner or later.

The illustrations of quartz implements, shown in the Drs. Sarasin's excellent publication, are beyond praise for the skilful manner in which the internal characters have been displayed. They would, perhaps, have been understood still more clearly had sectional views of the more important stones been appended. Several glass flakes are included in their illustrations. When it is remembered that rude glass razors, fashioned from the bases of beer bottles, were employed by our coolies some thirty years ago, one feels interested in the thought that Veddahs may have been chipping quartz, in Ceylon, at the same moment that Tamil coolies from Southern India were shaving with bottle flakes, just beyond their forest boundaries.

Dr. Seligmann, who was sent out from England to work up our Veddah lore (and I think that he has done so for all time), made some excavations in the Eastern Province which he did not consider to be very successful. The percentage of implements found in the caves was very small in proportion to the 'spalls' or refuse chips. As an explanation, I would suggest the probability that the ancients took refuge in the caves only during bad weather, or at night.

The author's thanks are due to Mrs. Robert Maclure and Mrs. Harper, who most kindly executed the coloured drawings for Plates I and II. These two plates have been reproduced most faithfully by the tricolour process, and bear testimony to the artistic character of the originals. In these and the remaining plates (which are from original drawings by the author), every stone is shown of its actual size.

I wish to acknowledge the assistance of Mr. E. E. Green, whose experience in the production of outline drawings for photo-process has been of great service to the writer.

Finally, it must be placed on record that the publication of the present work is largely due to the generous sympathy of Mr. John Ferguson, C.M.G., of the 'Ceylon Observer,' the only local paper that has taken a consistent interest in the subject.

CEYLON STONE IMPLEMENTS

FIRST PART

CHERT

THE FIRST TWO PLATES SHEWING THE NATURE OF THE TWO PALÆOLITHS FOUND IN THIS ISLAND.

CHERT.

IMPLEMENTS of Chert, with their accompanying "Spalls" and Chunks, in fairly formed flakes of every conceivable colour are found in Maskeliya district. Nuclei are rare and by this is meant the truncated conical cores which in our quartz series are so common. One small example is walnut-shaped and sized, and evenly worked, but as a rule they are coarse and seldom of the recognised form; for it appears that flakes were struck indiscriminately from any chunk at hand, most of the largest stones bear signs of this, and cores only appear when the lump is of superior quality, purer and more homogeneous in texture—and therefore more apt to flake as desired.

In this district (Maskeliya) Chert does not appear to have been the only material used, as silicious conglomerates also crop up here and there in flakes (3 instances, and a large Palæolithic-like chopper). The latter being the only example at present discovered. The nature of the Chert seems in many instances to have passed through some great chemical change, as the exterior surfaces have softened and become porous, for seeing that flint is, practically, of so imperishable a nature, one can only surmise that the geological character of the stone must embrace much variety in texture, unless indeed it demonstrates its own advancing age.

Chert to my knowledge has not been found in situ in this district (Maskeliya), though it occurs commonly in Matale; and in nodules in the dried-up bed of a stream to the N. W. of Segersta Estate, Puttalam. When broken open, these nodules are found to be hollow, the entire inner lining being encrusted with quartz crystals of precisely the same nature and size as those found on many of our

Maskeliya flakes, superficially, as if the "least resistance" in the flaking process had run along a vein of quartz in the chert. As a matter *àpropos*, our first Ceylon Chert flake was found on a sand bank formed as a talus to that same stream, one mile down stream (Fig. by Sarasin. Pl. 1, fig. 17).

The butt ends of our chert flakes assume almost the same character as those of our quartz implements, and the mark of the impact in flaking, is in many instances "an escutcheon" that proves its nature, birth and rank, for with almost a total absence of secondary chipping (retouches) any single chert flake might rank as a scraper, rechipping is very rare; in fact, as far as the present writer is aware, there is very much less sign of design in our chert flakes than is to be seen in some of our rudest quartz implements; of design in attempting to obtain some desired form or in making use of the resulting flake by chipping to a higher type.

Manufactories of chert implements are by no means common. In every instance of their discovery I have had brought to me the entire find of "spalls" (unserviceable chippings). In one instance the first chip flaked from the crust and its fragment, together with the flake resulting from the *second* blow, has been secured; these are now replaced (with cement) shewing the manner of the flaking, neither shew signs of having been used. May I mention that the only sign of the position of a "factory" lies in the presence of innumerable chips of all colours, many of which may have been serviceable if only required at the moment; apparently the artizan sat down on the spot where the implement was wanted and flaked till he got what he required. Chert has not the same rigidity and cutting power as quartz or otherwise many of the spalls might have been advanced a stage for some higher type than scraping.

Size of Implements.—Many of our chert spalls were doubtless made use of from about ¾ of an inch upwards as they shew signs of the resulting chipped edges from scraping. My largest flake weighs a little over 3 oz.; and measures 4¼″×2¼″ at the largest diameters. This is the largest chert flake in my collection. The average size of a serviceable flake I should put down as about 1½″ general diameter, the bulb of percussion well developed: a slight central ridge and an edge as sharp as the nature of the material will allow, a series of jasper and porcellanite presenting the keenest edges, and wherever there lay a probability that a small flake had been used as an implement, in nearly every case that flake was found of finer nature than that of the commoner kinds. A few two-ridged flakes and some very doubtful arrow heads almost complete the series of implements, were it not for some falcate flakes and a few thin spalls of a close-grained nature, some of which have apparent signs of having been used as knives, the disciform implements figured by the Sarasins being very rare in this district.

At the commencement of these notes reference is made to a large Palæolithic-typed chopper of silicious conglomerate (Pl. 2) found on the adjoining Estate of Glenugie 14-6-09. This was forwarded to Mr. Bruce Foote for inspection, and he was of opinion that the implement was undoubtedly an implement of the earlier type, but from its good state of preservation, the edges of chipping and general sharp angles of outline and surface he would not like to attribute to it the age which is generally allowed to implements of the palæolithic series, *e.g.*, Mr. Logan writes of a 400,000 years epoch of man and later authors hint at a million. Mr. E. Ernest Green to whom I forwarded the stone a short time ago returned it with the remark that the stone was not entire, but represented to him a fragment of a much larger implement and forwarded a diagram shewing the probable position of the last portion. Owing to the neat manner in which this stone fitted the hand for chopping, and the evident smooth surfaces over which the fingers must have strayed. I was disinclined to agree with him, I am afraid from a false feeling, that this might detract from the value. But on the first day of this present year 1911 two other stones were found, on Annandale Estate, adjoining this, about one and one-half miles in a beeline from the Glenugie find, and the condition of these stones leads me to acknowledge that I now believe that Mr. Green was correct in his surmise. Neither of these stones was entire, the larger had lost its point and the smaller represented the central portion only with about two and three quarters of an inch of the lower cutting edge broken away through a weakness of the stone, a quartz encrustment in an internally concealed hollow, but both stones showed no token of age by rounded angles and smoother surface or discolourment. The rock might have been fractured the day previous to their discovery, so clean and sharp are all their lines and faces.

These implements ought to command some respect from the educated natives of Ceylon for the reason that they are attempting to open up the history of their Island from a date backwards earlier than any of their archives afford us. At least we may assume so. For even if archæologists assign no great age to these stones on account of their apparent freshness of character, they must still mark the age of the palæo-neolithic overlap (in Ceylon) before the old types had entirely disappeared. The Glenugie implement was found amongst the tea on the top of a hill large enough to hold a small "settlement" or colony. The Annandale stones were found on a very small ridge in the tea, on an ordinary gravelly surface, soil open to both monsoon exposures, the ridge being about 30 yards in length with insufficient space for any "settlement," and the stones were found about 30 feet apart. In the Annandale instance, a few chips and quartz flakes lay on the scarp of the ridge, but neither at Annandale nor at Glenugie were there any chert flakes or chips.

Mr. Bruce Foote, in remarking upon the character of the Glenugie stone, wrote that he had found stones in India preserved in a similar state by the soil in which the implement had been buried, and added that the composition of the stone was itself remarkable in that he had seen none like it before in nature. But our three Ceylon stones were found on "the surface" and not in gravel pits, as was the case in Madras; they have had therefore all of exposure that could have been imagined in their case, an uncertain "æon" in the forest and nearly forty years in the open; possibly they have lain washed from all soil and exposed to the action of monsoon and sun for the whole of this latter period, and in addition to this an exposure during their rest in the forest, to all the acids, etc., of decaying growth, and yet they are fresh and sharp in their characters. Whatever the decision may be as to their approximate ages, which in any case must be indefinite, we have at last secured an intimation *that the neolithic age was not the only one known to our Ceylon ancients.*

Both of these last stones are of brown chert. On one, the original crust of the rock occupies about one-third of one surface, the obverse being comparatively flattened, less convex with sharp angles and edges. The smaller fragment is less homogeneous and is encrusted with minute quartz clusters which seem to penetrate the rock, and owing to this characteristic may be attributed the damage done to the implement. They are similar in general features to the drift and cave implements of England. The present sizes are No. 1, $4\frac{7}{8}'' \times 3\frac{5}{8}''$ from butt to apex. This stone may have been $6\frac{3}{8}''$ when entire in length; No. 2 measures $3\frac{1}{16}'' \times 3\frac{6}{8}''$ from butt to apex.

This last stone may have been $6\frac{6}{8}''$ when entire in length; one of these stones is figured on plate I.

The Drs. Sarasin asserted that our quartz series were all palæolithic in design of the 3rd period of the Magdalenian age of G. de Mortillet. The overlap of the periods probably having been suggested by the differentiation of the types, and that the absence of the polished celt was accounted for from the fact that the ancients passed over the neolithic stage (the characteristic features of which was the presence of polished celts and implements) from this phase of palæolithic man to the iron stage, which metal had probably been introduced by the Sinhalese or some other advanced race from the Indian continent. If this theory is correct, we need not hope to find any stone axe connecting the two later periods, and although we have not yet found any quartz implements of the palæolithic type, of the nature and size of those just described, we have in our collection implements of white quartz and opaline where the characters of the palæolithic chippings, the alternate blow from right to left forming the zigzag that maintains the edge, are most distinctly repeated. Our collections are still young, but up to the present time there seems to me to be strong evidence that chert and its companions bore the brunt of the manufacture of all the implements of our earlier Ceylon stone-age to the exclusion of quartz. On one side, our chert series, if judged by the European standard, are of all the most crude, scant of form, and, except in the rarer varieties of chert, almost destitute of edges; the ideas of the people, as far as culture was concerned, not rising above chopping and scraping. On the other hand, these ancient people seem to carry on the palæolithic type for awhile (in quartz) shewing design and wonderful progress in the manufacture of some of the most delicate tools and instruments imaginable. There is no design in any chert-flake that I have examined for comfort of the hand in gripping and pressing for the cutting motion of the fingers; whereas this feature is most noticeable in the quartz. The chert man seems to have chopped and scraped all his existence, for at present we have nothing that we can be certain of in the arrow head line; and the spatulate flakes which this valley has afforded make very poor knives. They may have been occupying the position of palæolithic man here whilst other lands were in the iron age as has been found elsewhere. So little has been discovered about our (Ceylon) stone-age that it is difficult to believe that there must lie "something" beyond our surface finds waiting to be brought to light; especially so, as we had been taught that Ceylon possessed none; nevertheless Mr. C. Hartley, a brother enthusiast, is pinning his faith to the Ratnapura gem pits for further developments.

Sir John Evans tabulates from a French author no less than about fifteen varieties of stone from which the common European celt has been chipped. Adding that the material most commonly in use in S. and E. Britain was flint derived from the chalk; but N. and W. not from any superior quality of the material, but owing to its accessibility, the metamorphic and eruptive rocks were more frequently employed; and he considered that in the palæolithic age all forms of these implements were much fewer and as a rule larger and *more rudely chipped.* This is exactly the state in which we find our chert series, which are so rude that no signs of progress appear to exist. Some few signs there are of attempts at rechipping, but, in proportion to the number of chert-flakes at present found, the instances are so rare, and the work so peculiar as to open up a wide world of doubt; existing marks may as justly be attributed to the occasional chippings of some too fine edge during the operations of scraping. There are no specially characterised implements such as arrow heads or straight concave and convex scrapers or knives, at least not in Maskeliya. "Points" there are, and unworked flakes, flat and fairly capable of cutting, but certainly not suggestive as such, except to a very sanguine collector.

Quartz, however, seems to commence where the chert men finished, and we have presented to us, arrow heads in many forms and of great beauty, worked on one side only; on one edge only; and in a few instances on both sides and both edges. Knives or cutting flakes of great delicacy of edge and finish, with rests for the forefinger, and for the grip of the thumb and second finger; the index finger being used for pressure. Scrapers for flat or round work; with roughened edges as if for sawing; for finer workmanship, graving tools, needles, and drilling points, and double pointed cutting tools of the most delicate nature; many of which are chipped evenly on one side for the comfort of the finger in pressure; and it is this last characteristic that draws us nearer to the latter end of the neolithic age erected mainly on account of the presence of polished stones, of which however none as yet, to my knowledge, have been discovered in Ceylon.

NOTE I.

Out of a large series of (chert) chips and spalls, debris of their factories, generally, speaking not 1% has been found with any patinations. An implement of the Disk nature was found by Mr. Preston Plumridge in my company, and this, originally of polished yellow-brown chert, shews by a later chipping that the entire stone has become patinated with a milky-white tone and polished surface. The stone was lying exposed on a small flat ridge on Maskeliya Estate. One other example might be mentioned. This, a very pale chocolate inclining to light-red, shews by some chipping subsequent to the date when first flaked from its nucleus, a polished rich salmon colour, as if this two-edged flake had been weathered for ages and at last "worked" (concavely and convexly). "Retouched" flakes of chert are very rare. In this last example the outer original skin of the flake appears rough through oxidation; the subsequent chippings shew polished surfaces.

NOTE II.

Out of a series of 230 *Quartz* "Points" and arrow-heads I have very few similar to those of chert; again the question of material dominating the form occurs. Arrow-heads (as we understand them) of chert are rare, but if we suggest the idea of "points," there are a few which might have been serviceable, as well as a very few spear heads, but these are all of the coarsest type. There are 12 of these "points" figured on plate V.

Their very uncouth forms almost suggest that the chert men had no use for bows and arrows, at the same time we must be just towards them and shew what we find; they might well have been serviceable, but whether they have or not, I leave the decision to those who may have seen such "points" in other collections.

NOTE III.

Since the above notes were written a very fine Disk implement of chert has been sent me by Mr. G. P. Cann of Kehelwatte Madulsima measuring $3'' \times 2\frac{2}{8}''$.

This implement is of intense value, as it makes a further attempt to bridge the chert-quartz periods. Two rather small implements of white quartz ($2\frac{2}{8}'' \times 1\frac{4}{8}''$) had been set aside for years, as, apparently, it was impossible to classify them; now, however, the implements can be recognised as Disk implements of quartz. The chert example is *broadly elliptical*, the two quartz stones are *narrowly* so. As collections become richer, it is possible that two distinct types of these implements will be found, the one elliptical as above, and at least one other more circular; examples of these latter exist in my own collection both in chert and in quartz. I should like, however, to see a series of this last before making a final decision.

My Chert Circular Disk measures $3'' \times 2\frac{7}{8}''$.

" Quartz do. do. do. $2\frac{2}{8}'' \times 2\frac{1}{16}''$.

Outlines of these might be given later on should the desire for further publications be sufficiently supported.

CHERT.

List of implements (or conjectured to be so) found to date.

Palæolithic, chopping implements.

Disk implements: probably used for chopping, Hammer stone for flaking, etc.

Scraping implements: of ordinary, single and double-ridged flakes, and flakes and spalls "useful" in great variety.

Scraping implements: of semi-elliptical form for cutting, etc.

Spatulate flakes: which may have been useful as knives.

Hollow scrapers: concave for stave-scraping, etc.

Points or arrow-heads: *none have been found stemmed or notched.*

NOTE IV.

Only one instance of chert was found in the Scarborough cave (see *Observer* 5-3-09). This has been figured by Dr. Seligmann (see Plate III, fig. 4).

NOTE V.

Only one instance of a chert core has been found shewing the fashion of flaking to obtain small flakes as in the quartz examples. This is of the size and shape of a walnut.

CHERT.

EXPLANATION OF PLATE I.

The stone figured represents the most perfect example yet discovered in Ceylon (as far as we know) of the older stone age; the palæolithic.

The discovery of two fractured examples on the same spot, on the Annandale Estate, allows us to come to the conclusion arrived at as regards the stone on Plate II; we now possess three specimens of the type of this age, all fractured.

This example is of olive green and brown chert; the hollow in the centre is a natural indentation on the surface crust of the stone. The edge as well as all the sides which shew the "work" are sharp, and the general appearance of the stone is fresh and recent.

Found on Annandale, 1st January 1911.

Del et Pinx Mrs R. Maclure.

CHERT.

EXPLANATION OF PLATE II.

The stone figured is the first of the palæolithic series discovered in Ceylon, as far as we are aware. Mr. Bruce Foote's opinion was that its greatest interest lay in the material, which is rare; but he quite failed to guess how important its presence was to us (in Ceylon).

Mr. Green, our entomologist, solved the question of its raison d'etre, and I now think with him that it discloses the basal portion of a palæolithic implement whose features may possibly have been those of the dotted lines. The material is of silicious conglomerate.

Found on Glenugie, 14th June 1909.

Since these notes went to press Mr. J. S. Coates, our Principal Mineral Surveyor, has expressed his opinion that this stone requires no "extension" to suppose its original features as in his opinion it is a perfect implement.

CHERT. PLATE II.

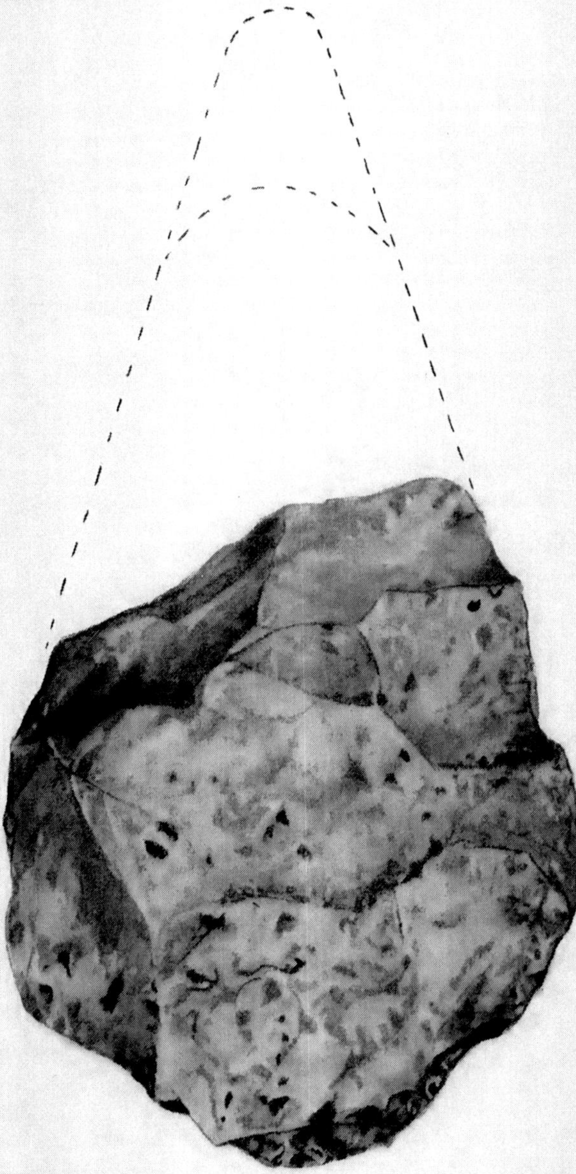

Del J. P. Pinx. Mrs. H. O. Harper.

CHERT.

EXPLANATION OF PLATE III.

Fig. 1. Cutting implement of rather unusual nature of white cherty material. *Mankulam N. Prov.*

" 2. The same style of implement in red chert, the butt end olive brown. *Maskeliya.*

" 3. The same implement in yellow brown chert of a compact nature. *Maskeliya.*

" 4. The same implement from the *Scarborough cave*, more falcate.

" 5. The same implement of different form and with, perhaps, a worked lower edge. *Gouravilla.*

" 6. Similar to No. 5, with an apparently worked lower edge. *Annandale.*

" 7. The same implement of different form in red chert. *Gangawatte.*

" 8. The same implement of common form in red chert. *Mahagalla.*

N. B.—Nos. 1, 2, 4, have sharp edges, not for use, on the upper sides opposite to the cutting edge, the remaining implements are all more or less purposely flattened to ease the forefinger in pressure.

CHERT

PLATE III.

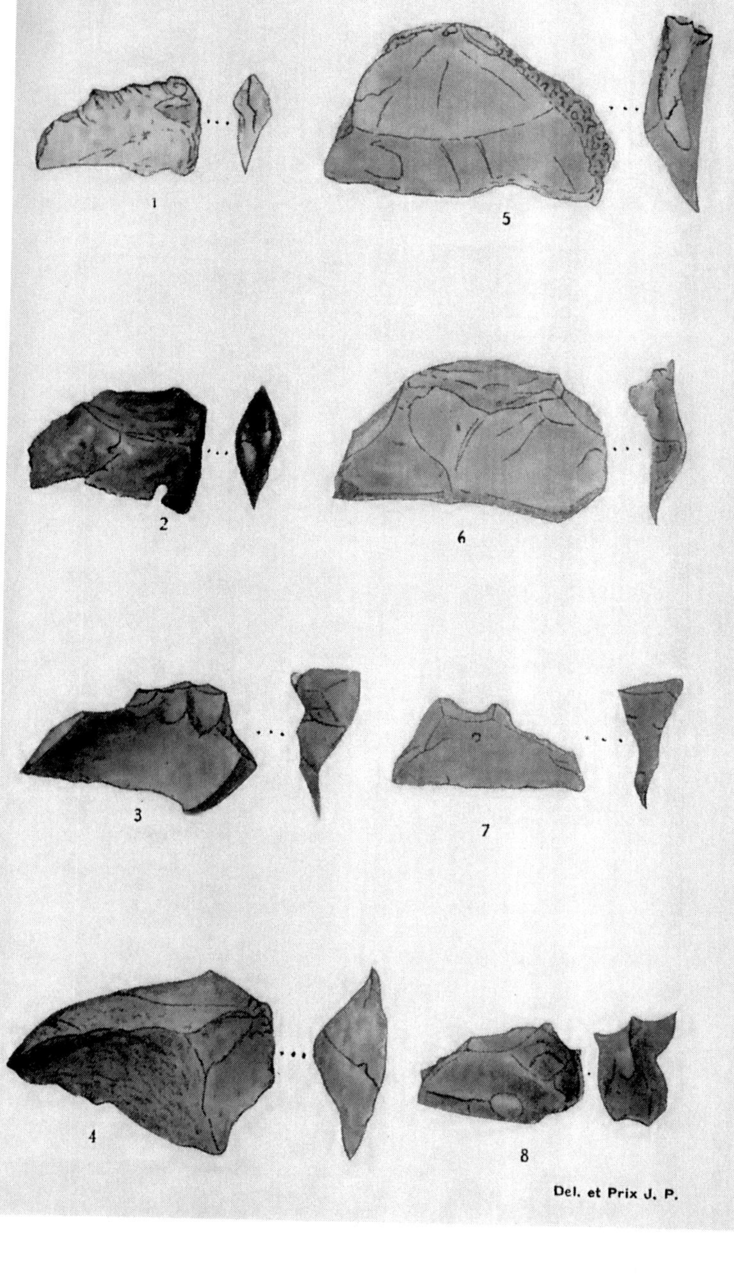

Del. et Prix J. P.

CHERT.

EXPLANATION OF PLATE IV.

Fig. 1. Small flake which has evidently been in use. *Merriacotta.*

" 2. Flake of yellow brown chert; butt end on the lower right portion; left side appears to have been "worked" to a cutting edge, which is fairly keen and was probably used as a scraper. *Brownlow.*

" 3. Flake of red chert very compact, cutting edge on the right, shews signs of having been used. The butt end is on the lower edge and shews the characteristic fracture at the point of impact with the flaking hammer, well. *Mahagalla.*

" 4. Implement of yellow brown chert probably used as a "scraper," hastate in form, and with the 3 edges all more or less sharp. Signs of use on the lower edge. *Cleveland.*

" 5. Cutting implement of yellow chert similar to the quartz example plate V, fig. 9, left edge very keen. *Gangawatte.*

CHERT. PLATE IV.

Del. et Pinx J. P.

CHERT

EXPLANATION OF PLATE V.

POINTS OR ARROW-HEADS.

Fig. 1. Brown chert flake, point missing, edges fairly keen; probably of use as arrow-head. *Stockholm.*

,, 2. Dark chocolate point, right rather flat, left edges keen, apparently quite serviceable. *Maskeliya.*

,, 3. Pale yellowish opaline chert; butt end on right; the yellow lines and patches denote massed quartz crystals, a serviceable point, from *Maskeliya.*

,, 4. Brown chert bulb of percussion, seems to have been used. *Gouravilla,*

,, 5. Very coarse single-ridged, bulbed flake, butt end a base. *Bargrove.*

,, 6. Pinkish yellow, or salmon-coloured, double-ridged flake, butt at left base. *Maskeliya.*

,, 7. Chert flake of yellow ochre, butt at left base, the only point with any form, in my collection. There are apparent signs on right and left edges of "work," but both are dull. *Gouravilla.*

,, 8. Single-ridged flake, butt at base, an appreciable point.

,, 9. Point formed from the bulb of percussion showing the hollow on obverse, natural white skin on the right surface, apparently has been "worked." *Brownlow.*

,, 10. Has been a serviceable point or arrow-head of some cherty conglomerate, point missing. *Annandale.*

,, 11. A point whose left side shows the entire bulb of percussion; the base is quite keen and is quite suggestive of the "scraper."

,, 12. A point of olive brown chert.

CHERT. PLATE V.

CEYLON STONE IMPLEMENTS

SECOND PART

QUARTZ

QUARTZ.

Quartz Implements from Maskeliya.

The material itself, at least, much of it, is of such inexpressible beauty and its characteristic features so dependent upon the amount of daylight filtered through it, as to render it a task of some difficulty to attempt to catch the delicate nature. The angles and curves of quartz as brought out by the prehistoric workmanship, added to the natural internal beauties in the stone, present very many beautiful features. The delicate lines of fractures, and fasciæ, of clouds, and minute spots of glittering dust, of snow-like flakes with their internal fissures, crossed on the exterior surface by the ribbed and fanlike streaks of conchoidal fractures brought about by the human hand in flaking and shaping, seem to give this material some very confusing characteristics, especially as all these markings seem to change their position according to the angle of light, markings discovered and figured at 6 A.M. frequently disappearing entirely at noon. From the Journal of the Royal Society of Arts, No. 3072, Vol. LIX, I steal this encomium. See Prof. Tatton's Lecture I "* * * *ut nulla id arte possit æquari*" and that SiO_2 is its chemical equivalent.

Arrow-heads.—There are not a few of the types of arrow-heads, which are so common in the English and European series, missing amongst our quartz forms—and in many, nothing but the notch at the base tells its history; this notch with a little chipping of an edge or point has formed the arrow-head.

The sections of these, in a large percentage of our present finds, assume every conceivable variation of the triangle, dominated clearly by the nature of the quartz rather than by any intentional blow to separate it from the mass.

The white quartz series is rough and irregular, that of the purer crystal is more uniform ; the former examples are rare, though pebble skins of leaf-form were not despised, and it is a fact that the only example of a long-stemmed arrow-head which we might presume to be their highest developed form, is of white quartz, the coarser stone. As the Ceylon example has already been figured by Dr. Seligmann it would seem hardly necessary to figure it here, but as it is at present unique for Ceylon, and makes our series more perfect it has been drawn, but without sections. See Pl. iv, fig. 2.

Sir John Evans remarks that in prehistoric days the arrow-head doubtless represented the highest form of the stone chipper's art. At least it is fair to assume this on the strength that there is more apparent design and actual work shown in the attainment in this branch of their artillery in Europe. But with the fine specimens of their workmanship before us, the intelligent grasp of the possibilities worked out in pure flint by those neolithic ancestors, who had a most discriminating love of the symmetrical, and intense surprise at what was found useful, with the undoubted discoveries that the same usefulness was found here, one cannot refrain from doubting that all our forms of this weapon were fashioned from intentional fractures ; we can quite understand a people working in flint for a few years and attaining marvellous dexterity in so workable a material, but one would much doubt whether the same people would have attained any greater proficiency in the craft than that shown by our Ceylon ancients (remark *viva voce* by Mr. Bruce Foote). Some of the arrow-heads which have been secured are of so gross a shape and dimensions, whilst others are of so beautifully elegant an outline, and both the former and latter series so free from secondary chipping (retouches) as to assure us that the original flake was accidental.

Drawings of quartz cannot adequately express the beauties of the stone, the material, for it must be remembered that nearly every line which is represented in black in reality appears "inside the stone" as *white* on a darker ground. Herein lies the charm of the Drs. Sarasin's work. Nor does the shape of the implement always guarantee, at the first glance, the nature of its attributes, for it seems fairly evident that no truly useful shard or flake was discarded. For example, some of these drawings might very reasonably be regarded as "arrow-heads" by their shape, but as the edge is the important factor, so this governs the use of the flake, and I should calculate that about 80% of all implements come under the category of cutters and scrapers or unworked implements—*see* Pl. v, fig. 4. Nor would the arrangement of the implements according to their shape or type produce a very satisfactory estimate of the actual capabilities of the worked stone, the individual artifact. For, if the implement is formed from the casual fracture, intentional, yet not always answering to the will of the artificer, many times the fractures from a mass of this quartz must have taken upon themselves a shape whose outward aspect proclaims them to be one implement, yet whose actual value lay in the being of some implement far remote.

The stones are all drawn to their actual size and many of those figured have already been exhibited before the R. A. S. (Ceylon Branch); these may be recognised as follows :—

Quartz Pl. i, figs. 1, 5, 6 ; Pl. ii, fig. 2 ; Pl. iii, fig. 6 ; Pl. iv, figs. 2, 3 ; Pl. v, figs. 3, 4. I am indebted to Mr. Preston Plumridge for some nice examples from Midlothian Estate, distant about 3 miles, some of them of great beauty, and to Mr. William Vaughan of Madulsima for some characteristic stones from that district ; and to some Singalese school girls on Clodagh Estate, Matale, for some nice microliths ; to Mr. Bainbrigge Fletcher of the "Sealark" for one good core and one good flake from Hambantota ; to Mr. Thos. Farr for a small selection of flakes and implements from the Horton Plains. Most interesting these last, as recording flakes from the highest elevation at present obtained ; for previously, I had had sent me some chips of stones from the Horton Plains which I was doubtful about, and when the Drs. Sarasin subsequently dubbed them "dubiose," I felt greatly relieved as I had pained my friend the sender by a similar remark. Mr. Farr remarks that all the "workshops" on the Horton Plains and Ambawella patanas are found on the roads only, *where those roads have cut them through*, and there only noticeable.

MEMORANDA.

"Then Zipporah took a sharp "stone" Ex. 4, v. 25 "stone" also "knife." And they cried aloud, and cut themselves after their manner with "knives" and "lancets"—I Kings, XVIII, v. 28. Flint Flakes, according to R. A. Stuart Macalister in his Bible Side Lights, pages 98 and 103, were in use in Geza up to the time of the Captivity. Cir. 588 B. C. And there is no doubt whatever that in the two scriptural references given above, the common flake was made use of. "Perhaps at this time handed down from the ages as a part of their mysterious ritual necessary adjuncts to their solemn occasions" (Sir John Evans). Palæolithic chipped stones were discovered in India in 1863 by Mr. Bruce Foote; our Ceylon Neolithic, at last recognised as such, were first discovered in 1884. Our first Ceylon Palæolith, if recognised as such, 14th June 1909.

CEYLON PREHISTORIC STONES—NEOLITHIC.

There seems to be very decided opinion that our Prehistoric stones belong to the Neolithic series. Mr. Bruce Foote, our only Eastern authority in these matters, opined this early, and later our British Museum authority (Mr. C. H. Read of the B. M.) who had only seen some rough sketches endorses this opinion. A slight reference to the history of this matter might not be out of place here. In 1884, Mr. E. E. Green, our Government Entomologist, and the writer, found stones of undoubted human origin in Pundaloya (3,000 ft. to 4,500 ft.)

and Navalapitiya (2,000 ft.) respectively, both of them collecting as occasion offered and endeavouring to enlist others to this same conclusion without result, and, in fact, no interest in the subject was shewn, as far as the absence of all printed matter is concerned, until the Drs. Sarasin who had, on some three occasions, visited the Island and who were working at the anthropology of the Veddahs, made a most interesting discovery in several of the Veddah caves in the Eastern Province, of stones chipped and worked, of evident human origin; and subsequently found them scattered over the ridges and grass lands (patanas) of Uva. They had previously been found in similar positions by the writer, on patanas and opened lands at Navalapitiya and by Mr. E. E. Green in Pundaloya. The Drs. Sarasin at once attributed the source of these stones to the ancestors of the Veddahs.

From many sources we are assured that the series are utterly unlike the Neolithic Flakes of other lands; not so much from their form as from the material of which the greater portion of them is composed, a quartz of an ice-like purity seems to have been the favourite variety, although the cores, many of them of very great beauty, show that they were not slow to make use of any variety in this same quartz or rock crystal; and the forms of the implements obtained from crystal of the toughest and most obstinate nature has called forth some eulogy from the pen of Mr. Bruce Foote, to be alluded to further on.

In very many instances the intentions of the manufacturers are difficult to fathom; the only means we possess to arrive at a fair conclusion is, our very common sense of what their desire was, or the necessities they were striving to supply.

We have been assured by the Drs. Sarasin that the Veddahs are a small handed race, and if these ancients were their ancestors, as certain it is that their implements were of a small character; for the majority of the implements at present discovered are much below the standard, with the exception of the Pigmy stones recently discovered in England.

The larger and coarser flakes are many of them without any sign of workmanship, if we except that portion of the stone which was removed to form a rest for the forefinger; on many of the most useful chips, indeed, it is the presence of such a "removal" which seems to prove the intention of the manufacturer, and the position of this space for the forefinger, at once seemingly declares the use of the implement though there may be no other sign on the stone. The flake comes to hand most naturally and conveniently for skinning or dividing flesh, etc., into portions, for horizontal or perpendicular use, see Quartz Pl. i, figs. 1 2, 5; Pl. v, fig. 4. The nature of the material precludes, in the majority of these coarser flakes, any chance of assurance that the edge is anything else but a natural fracture, but the presence of a forefinger grip at once appears to declare the nature and genuineness of the implement and allowing a decision to be formed at once whether the implement was intended or not for service.

These remarks apply to the microliths as well as to those of a coarser nature; for some of this series are beautifully worked on the side opposite to the cutting or working edge for the comfort of the grip; whilst others show no sign whatever, wherever the fractures have fallen convenient; more especially is this noticeable on the pellucid stones where "work" of any nature would be distinctly visible. Should the material have fractured conveniently, little or no "work" was expended, Pl. i, fig. 6. In fact, the aim of the ancient people was to get an edge; having secured this, they went no further, if the intended implement could be held sufficiently firmly for work required.

The majority of the Implements consists of scrapers straight, curved (concave or convex), broad or elongated, even or serrated "edges," and this description refers also to the microliths to make which it seems that only the toughest portions of the material were requisitioned, though on occasions, the pointed crystal is found to be worked on two sides for the finger grip and the point straight-edged, rechipped to sharpen it, and implements as small as ⅜" appear to have been thus worked.

The habits of the manufacturers, their manner of living and their desires expressed in the forms of these stones, conjecture alone can help us in realizing. (Their implements found in the caves are precisely similar to those found in the open). As we have had no detailed description given us by the Drs. Sarasin or as yet by Dr. Seligmann, it would be interesting to learn whether the microlithic series had also been found in the caves they searched. At any rate we may surmise from the presence of the stones that these working prehistorics took to the caves in bad weather, and that they carried their shelters with them, in the form of skins, when they roamed farther a-field. In fact, that they, probably, sheltered themselves from wind and rain with the skins of their quarry when moving about; more with a view of the shelter they gave at night than by day; for so slightly do the Veddahs of the present day clothe themselves that it can hardly be said that the idea of wearing clothing was evolved from the habit of carrying their night gear with them.

I opine that these hunters watched the ridges where the tracks of the game crossed them, for water; stunned the game with ordinary stones, or, wounding them, ran them down and killed them with heavy sticks, perhaps using a chipped stone as a "misericordia!" In fact, the idea strikes one that they worked in concert, in numbers (the heaps of chippings from their implements suggest this), that they camped on the ridges to intercept game crossing them for water, and at a given signal combined for a running hunt; it is very difficult to imagine any other manner, for the only examples of arrow-heads we have found are of so uncouth a character that I strongly doubt if these men counted much on their aid. Mr. Thos. Farr, our modern "Nimrod," writes: "What I cannot understand is how any implements or arrow-head made of quartz could be heavy enough and sharp enough to penetrate the skin of a living deer." Doubtless they were of use for birds and small game only. As year by year the same spots were visited at the change of the monsoons when the game shifted their feeding grounds, so these hunters followed, choosing very much the same spots for their leafy shelters, "something that no animal would take notice of" (Dr. Sarasin), manufactured their implements when required, on the same spots, and dropped them on the same grounds. This might account for so many being found on the ridges. They sought their food on these ridges and carried the stone, manufacturing their requirements on the spot.

I am not of opinion that the material was found on the ridges. Nor do I imagine that the stones they made use of as missiles were "worked" in any degree, though they may have *selected* them from the nearest stream. I take it, that distance was of no moment; if only game was abundant, there they carried their outfit, probably a "favourite" proved-hard-stone, for flaking; and a few cutting stones, for immediate necessity, to be thrown aside when that particular season was over.

Mr. A. C. Logan, I.C.S., page 66, seems to find that *absence of material* led to a district being absolutely uninhabited by the quartzite people of India, may it not have been scarcity of game? or water? or difficulties in the hunt? Anyhow, in Ceylon it appears that the material was carried great distances and the implements manufactured on their camping grounds. The late Mr. James Parsons, our Government Geologist, seemed to think that the material, or some of it, was obtained "in situ" from cores and veins in the coarser quartz rocks, though I have seen nothing of that nature in the ordinary occupations of draining and road making, building, etc., on an estate, with fully 40 years' experience.

Situations (in Maskeliya and Navalapitiya).—As previously stated, the stones are found on every available point of vantage, on the ridges principally, and the presence of a seeming "piece of glass" is quite sufficient guarantee that there is more about; and although single implements may be picked up here and there on the flats, the manufactories are always on the ridges, where implements of every description may be gathered. Only like mushrooms, they must be looked for! Many of these manufactories retain stones of great beauty, from the discarded, glittering, core to the delicate little graving tool, worked along the whole surface of the back in a surprisingly artistic fashion, and with a point as keen now as the day it was formed, *perhaps* 2000 years ago. (*See* Pl. III, fig. 11.)

The date or age of these implements must have been fairly recent, for, arguing on the lines of the general hypothesis that soil is washed away in open land at the rate of two feet in a century, and allowing that nine inches has disappeared since the forest was felled thirty-eight years ago, we are still confronted with the fact that when the forest, after being felled three months, is burnt off, these stones are found, at once, on the surface. The estate adjoining this felled and burnt off fifty acres this year 1908 about March; many very delicately formed stones were found, at once, on the main ridge. (They are probably lying over the ridges in every mountain in the country whether cultivated, abandoned, virgin forest or grass land.) Possibly 3 to 4 inches of fern roots, fibres, and leaf mould were burnt up on the occasion stated above; to this add 9 inches for 38 years' exposure to the monsoons and the result is the same.

Beside my own implements, I have seen only 4 stones; these had been dug out from the Veddah Caves by Dr. Seligmann. The Sarasin finds were not available, nor, at the time of my visit, in September 1908, were there any exhibited in the Colombo Museum. I am at a loss, therefore, to make any statement of the limit of size of the implements discovered elsewhere up to date. My largest stone (Pl. I, fig. 2) measures $2\frac{11}{16}$ inches. My smallest $\frac{6}{16}$ inches (Pl. III, fig. 5), the former a coarse cutter (for flesh work), the latter I should presume microlithic for cutting and engraving or ornamenting bows, arrows or weapon handles. The larger stone (Pl. I, fig. 2) is a pretty rose-tinted flake with finger rest; the smaller stone is interestingly worked on back: of this we may be assured with regard to all our implements, no great *depth* of soil has been formed since their connection with man. Naturally, it might be argued that the mould formed on the ridges would be the earliest washed away by the rains, but the fibrous roots of the forest and those of ferns would retain the stones; and this probably is the case in many instances, but in the open land, on flat ridges, undoubtedly, nine inches of fibrous matter and mould must have been carried away in 38 years, and even here the stones lie on the surface.

The localities where these implements have been found are worth recording, Hambantota the pure crystal, Matale, Uva Pusselawa, Ouva, Madulsima and Diyatalawa, Newara Elliya, Dimbula, Dickoya, Bogawantalawa, Balangoda and Maskeliya, Horton Plains, Galle and Mankulam N. P., and one correspondent writing from Rangalla declares that there are no signs of these stones in that district, probably the country did not lend itself to successful hunting either of Deer and Game by the Prehistorics, or of Prehistoric stones by the present man.

Dr. Seligmann informed me that with much labour he obtained about 300 chippings in caves, all of which when washed out turned out about 4 per cent. of worked implements. On these ridges it would be labour indeed to count over the possible stones, even if they were removed for the purpose, nor have I attempted the census. See note V appended. The range in the dimensions of these implements $\frac{5}{16}$″ to $2\frac{11}{16}$″ is great, and this, I think, may imply two sections of their communities; those who hunted and skinned, and these made up the majority, and those who remained in camp, and did work which concerned the hunters as well as the rest of their clans Domestic. The clumsy arrow-heads and fleshing implements are easily separated from the delicate cutting, graving and boring tools, as much from their size as from the roughness of their manufacture. Yet the knife of the Deer Skinner and the Phlegm of the Surgeon might well have found a place in their workshops, in the knife, arrow-head and lancet formed from these limpid crystals.

It is most difficult to imagine that they were all, both small and large implements, made from design, but rather from the splinters of "lumps" and "chunks," having a natural tendency to "come away" in serviceable forms. That instances of design were few I can easily imagine. For why should the child starve whilst the flesh provider "polished" his implements? and when it can be demonstrated at this present day how difficult it must have been for these really clever quartz splitters, to arrive at any conclusion as to form, I think we may exclude ornamental details which, in the working of this material, would in this present day show a strong tendency of madness.

The Scraper we may regard as the unworked implement, the convenient shard or flake, and any flake may have been applied to this purpose, without having on it any mark of manufacture. In many instances the edge is purely a natural fracture, and the "work" appears in a space suitable for resting the forefinger Pl. I, fig. 1, for grip or pressure. For some time I was completely puzzled by the appearance on one side of an edge as if the edge had been partially given by attrition on sand, but the riddle is immediately solved by an attempt to "flake" the "roughness" in the manner we suppose that the ancients employed; except by continuous training of the muscles of the arms we never could attain their dexterity.

The right arm must have become mechanical by practice enabling them to remove most delicate surfaces in their "swing" and those movements carried out with a "*rasping*" stroke, undoubtedly formed the lines of scratches on the edges of many of their keenest implements. Mr. James Parsons, our late Government Geologist, demonstrated the principle to me with great success, and the implements he then developed will, I trust, find a space in our Colombo

Museum. The conclusion I think that will be arrived at is that the implement to be manufactured first displayed in the fragment some characteristic desirable, then, if the edge was good, the formation of the grip was attended to ; in the case of large implements a proportionate seat was made for the forefinger, in the case of the small implement, a distinct rough surface was formed to the side or sides opposite to the cutting edge, and, if required, portions were chipped from the two end surfaces for the thumb and second finger grip whilst the forefinger was used for pressure. Many stones of the microlithic type show signs of hard work in a neat and comfortable rounding off of the abrupt edges (Quartz Pl. III, figs. 1, 2, 4, 5, 7, 8, 9, 10, 11).

Mr. Bruce Foote in his Palæoliths of S. India, page 22, remarks on the apparent usefulness of the unworked implement. "Flakes possessing clean cutting edges, being the simplest form of implement and produced whilst forming some implement of higher design, have many evidences of having been found of convenient use" and for variation in form he quotes Evans as depending "on the defects of the flints." In any view taken of the value of an implement the main *raison d'être* is the edge or point.

Flaking.

This process has been explained in various ways ; some hold that chippings were made from the inside outwards, this could hardly have been the case with our Ceylon quartz. For Mr. Parsons assured himself, in my presence, that the blow in the majority of instances acts better when struck from the outside inwards. Naturally a hump or irregularity existent in the centre of some implement to be formed, might be chipped from any angle and in any direction. (We have the flaking implements but none are figured.) Mr. Logan surmises that the rudeness of the form in the Indian stones is largely due to the material, and Mr. Bruce Foote gives our Ceylon originals all credit for their persistence and pluck in attempting to form an implement from such obstinately tough material. Therefore it seems that material dominates the form, and it is hardly surprising that our Ceylon Implements present a facies so very different to that of any known. It must still be remarked that where good material had split to useful dimensions the form of other lands can be recognised; specially is this so in the microlithic series, where Mr. Bruce Foote recognised at once the conventional graving tool (Pl. III, fig. 11) and the cutter (Pl. III, fig. 10) very well known to Dr. Seligmann.

So very few people, who are interested in this subject, have the time to spare, to hunt up the various districts and collect instances of divergence in type and in the selection of material, that it is difficult at this stage to point to any progressive development, so little has been done, so restricted in quantity of "material" collected as to make any decision on the subject almost worthless.

That there has been progress we have a right to emphasize, for no one can examine the chert and the quartz scrapers; and the microlithic graving tools and hooked cutters without a conviction that knowledge had increased. Although the march of intellect has left such fairly formed ideas crystallized in its track, the present age man has declared himself unable to demonstrate for what special purpose certain implements were formed, although he is fairly able to decide in what manner these stones were worked.

A list of the various implements is given which by no means would satisfy the present-day collector. We do not know what these implements were used for, but can only guess and give them a name which may demonstrate the nature of the work we think they were intended for.

I would call attention to a very interesting fact. Amongst the spalls and *débris* of "camping grounds" I have picked up crystals which have been detached from their original formations and apparently have been made use of. The points are worn, and the shafts for some $\frac{6}{8}$ of an inch present the appearance of having been used both as punches and as borers. As regards the "punch;" if these men were capable of appreciating such an instrument, I can point out only two conjectural instances: two stones have come into my possession, the one an ordinary core, the other a possible implement. In the former is a furrowed groove, impossible of structure otherwise than by a punch; the other is a possible implement, hooked cutter, or some "thing" *formed* with a central hollow of the exact contour and outline as if some hexagonal crystal had been forced through the stone. This is, as I say, purely conjectural. Should the use of this feature in construction be proved, it may surely bring the age-limit of these "implements" down to a date when the iron-tipped arrow worked alongside of its less enviable chum And I have in my collection such an implement, an ordinary quartz crystal, which may have carried out such operations. The two ends are smashed as if the impact had been given by the hammer (by a third stone).

The tridentate natures (Pl. V, fig. 3) given to these implements is due to the fact that two blows were given to the intended implement by "the flaker" cross-wise *over* and *downwards,* thus forming three teeth which may be counted as the "hall-mark" of many a stone which would not otherwise be called a prehistoric implement. Tridentate stones are common.

The tridentate stage may have been evolved from the very ordinary course pursued in chipping away unnecessary humps and excrescences from the implements they were fashioning, for the forward and downward blow, of one stone over the other, produces this effect. It may then only have been discovered, that, given a hard and rough stone, the attrition of other matter was easier than by means of a hard and smooth stone. And a kind of "excoriator" was introduced to wear away surfaces to a more even plane. If not for this, it is hard to imagine where reason "held her sway" for a very large percentage of stones (I won't say implements) possesses this feature of two gashes made in the stone.

size of stone immaterial, resulting in the tridentate form (Pl. V, fig. 3).

Special implements were used for flaking: and apparently the flaking implements were assorted as to the sharpness or otherwise of their working edges, possibly the different natures (hardnesses) of the stone required selected tools.

For these tools have their working edges at any angle from 60° to 70°. And (as pressure is the generally accepted theory) we can conceive of an edge helped to some convenient keenness by the aid of the implements we have discovered. A sharp-edged stone backed by a short wooden baton pressed against uneven surfaces removes excrescent roughness, leaving a trail of hollows in the line of the least resistance; and the purer and more homogeneous the material, the greater the length of the furrows, and the more even the surface of the work. Our best example appears to be one of a limpid crystal and is figured by Dr. Seligmann. It is probable that these implements were used for working *from the edge inwards;* if otherwise, it might tend to spoil the edges of the implement to be fashioned.

There appear very few instances of continuously hard work applied to these flaking operations, most of the implements having the appearance of the natural fracture.

It may be objected that many of the stones depicted here are nothing but "shards" and "chippings;" in very truth they may be, but if they are compared with others of the same type and actual form, it will be seen that nothing but a few chippings and "retouchings" separate them from the actual implement; they are, in fact, the unfinished implement, and are of value to demonstrate how insignificant an implement may be or how significant and fine the line of difference can be drawn, between "worked" and "unworked" implements. It must be allowed, that the most important implement amongst these men was probably the knife, that no human agency, at that period amongst this particular race, could produce an edge comparable to that which was formed by a natural fracture, that the most they could do was to improve this edge occasionally, and that on those occasions they never attempted one with a low angle. Some implements with a cutting face for stone work, "Flakers," chiefly have the planes of their surfaces inclined to 67°. If the edge flaked suitably, they appear to have made the grip comfortable, or not, as it suited them at the moment.

If the history of our Neolithic men were thrown back far into the ages, there would be a certain reason for finding so many flakes in the open and few in the Rock shelters which do not seem to be common up in these hills, but our first Maskeliya Cave certainly yielded well for its size. Mr. A. C. Logan, I.C.S., in his brochure on the Old Chipped Stones of India, page 50, estimates 400,000 years between the present time and one of the South Indian Palæoliths, and hints of possible "traces of man" 6 or 700,000 years ago, page 73. (The latest in the field hints the million).

In the manufacture of many of the crystal flakes, the design, manner, and intention of the manufacturer seems undoubtedly to have been regulated by the behaviour of his first chippings of the pebble. This is chiefly observable amongst the Microliths. The delicate finish could hardly have been given after separation of the flake, but previously. It is difficult to conceive the operation carried out otherwise. I am presuming that the pebble having been split in half was subjected to a hammering to remove the two rough surfaces then flaked away in the usual manner, each flake being more or less half finished by the previous hammering, then finished off very gradually. I have these small implements in many sizes and the general workmanship on the back, the smoothing for the finger grip invariably leads me to this conclusion, which after all is of the simplest and therefore more open to discussion.

That these men were not particular as to the material they made use of we are assured by the discovery of a core of Calcedonic quartz of which no flakes have as yet been found, and of an ordinary notched scraper of tourmaline which in the outline of its useful edge differs in no point from their very common scrapers of white and pure crystal quartz. This scraper has a very well worked edge of about 80° and seemingly is about as clumsy a fragment of rock as one could well imagine. The work on it, however, is of undoubted character, and the difficulties in its production can well be imagined when it is known that tourmaline is of so hard a nature that its powder is made use of to grind and polish gems. Its longest diameter $2\frac{1}{8}''$, breadth $1\frac{1}{16}''$. It has apparently been flaked from a larger mass in the custom usually observed in quartz, is of black tourmaline and was found on Blair Avon Estate by Mr. Preston Plumridge who has kindly added this to my collection. See Pl. II, fig. 3, probably used for skinning staves, etc.

Iron Ore.—I have also in my collections a flake of iron ore, fractured and flaked from the original mass in a manner apparently precisely similar to that employed in flaking chert and is seemingly an implement of the concave scraper type.

QUARTZ.

List of Implements found to Date.

Hammer Stones; rounded pebbles of quartz rare in Maskeliya.

Disk Implements; elliptical choppers rare in Maskeliya.

Scrapers: 1 {Concave / Hollow} smooth; 2 {Concave / Hollow} rough; 3 convex in various forms; Dentate in various forms; straight edged with guides; Double curved edged, with high and low angles.

Cutting Implements; in various forms; pointed for boring or piercing skins, etc., for graving handles of tools, etc.; microlithic "cutters" of many descriptions and "scrapers."

Flaking Implements, with very low-angled edges.

Arrow-heads and points; *with tang or stem;* without stem and without notches; with single and double notches; leaf-shaped and triangular, and many useful forms.

Sawing Implements; with regular and irregular teeth.

Boring Implements; similar to European form

NUCLEI, COLOURS OF.

Calcedonic quartz; opaline, and pale cinnamon, pale yellow, white, pure ice-like crystal, smoky olivine, rosy, black tourmaline.

NOTE I.

An interesting circumstance connected with the finding of a white quartz flake might be recorded. A much weathered and rolled pebble of diaphanous quartz from which a flake of 2½″ had been struck, was found on Maskeliya Estate, 14th November 1909. The flake which had been struck from this had been brought to me a year previously from Gouravilla Estate. The two estates are two and a half miles apart.

NOTE II.

Dr. Seligmann has copied for his book on the Veddahs, Fig. 2, Plate IV.

NOTE III.

The Drs. Sarasin's Figure in Plate III, No. 89. an implement which may be a larger form of microlith, but it seems it is, interestingly, the only microlithic form in their whole collection. Sir John Evans, in his ancient Stone Implements of Great Britain, page 325 of the Second Edition, mentions and figures three microliths from the Vindhya Hills of India, and states that "similar implements have been found in Egypt and Belgium," remarking that "Such an identity of form at places geographically so remote does not imply any actual communication between those who made the tools, but merely shows that some of the requirements of daily life and the means at command for fulfilling them being the same, tools of the same character have been developed irrespective of time or space," and on page 276 remarks on the chance of a punch having been made use of as follows:—"I have already suggested that in striking off such small flakes some sort of punch may have been used instead of the blows being administered directly by a hammer."

NOTE IV.

This is written with regard to the "working" of small cores. Sir John Evans mentions that when these "cores" or nuclei were of elongated proportions, they were frequently turned to service as chisels or mining tools. It is extremely difficult to understand why our minute Ceylon nuclei should have been worked in so delicate a fashion, producing flakes so thin and small and evidently of so useless a nature. Since reading Sir John Evans' hint on the apparent existence of a punch amongst our English prehistorics, for getting away the minute flakes, it seemed to me that the object must have been to obtain flakes and not merely to round off an unserviceable core, and that our Ceylon cores were the result of mechanical "smashes," blows dealt by the swinging right hand on a stone held in the left, and, by an expert, thus wonderfully, regularly, distanced and proportioned, in the depth of the blow. The flakings may have been used, as surmised, for "Grinding," but I have not yet found a stone where the signs were clear enough to assure me that any "rubbing" or grinding had been practised. Instances where I believe a punch-like implement may have been used, exist in the many examples of obstinate fractures, when the "line" desired failed to appear during the flaking process, for quartz is of so intractable a nature that an intentional design seems rather the exception than the rule, and this instrument may have been used in easing off obstructions in an edge, or indeed in forming some specialised implement of the cutter or excoriator character. I have seen Mr. Parsons produce these effects, and have, indeed, myself imitated him, in forming the elongated furrows in the stone by the swinging smash of one stone against the other, just as Sir John Evans suggests. As regards the existence of cores, they are found everywhere wherever the flakes are found, and, in truth, they are always amongst the first signs of implements that are brought to me. Out of 120 well-worked nuclei my largest (No. 2547) measures $1\frac{10}{16}''$ × $1\frac{5}{16}''$, my smallest (No. 2676) $\frac{9}{16}''$ × $\frac{7}{16}''$. The former is pear-shaped, the latter conical. Many of them are of great beauty, and it seems almost certain that these small nuclei *of good splitting stone* were carried about by these ancients, and small cutting flakes struck from them as occasion demanded. For flakes from this crystal possess razor-like edges, and must have proved serviceable for many purposes of the chase as well as for domestic use. In a collection of over 3,000 implements and flakes, I have not found one where the practice of rubbing or grinding can be definitely stated as being carried out, nor have I come across any attempts at polishing. The backs of the microliths *may have been rubbed* with cróss-cutting stones.

NOTE V.

In connection with the percentage of implements found amongst the chippings, and previously referred to, I was unable to answer Dr. Seligmann's query as I had really taken no census; now, however, some idea can be given in the matter. In November last two lots of flakes were forwarded to me from Matale East; in one lot 381 flakes yielded one doubtful implement, in the second lot 507 flakes produced 9 well-formed implements and 37 ordinary scrapers, in the aggregate 888 flakings produced 47 implements, say 5%.

Dr. Seligmann found *in the caves* 3%. Naturally, *in the field* the percentages would be less. I have made no attempt to discover the proportion in Maskeliya. The proportion of cores to useful flakes and implements, as represented by my entire collection, works out at 4%, but I must have refused twice as many in the course of selection, and I think it is worth recording that most of the nuclei at present discovered are the cores of limpid stones. And I do not recall ever having noticed white quartz nuclei of that nature from which the majority of coarser implements must have been struck. I am obliged to Mr. Hartley for 2 from Uva. The "white quartz" pebble is represented by cutters and scrapers, with a portion of the red brown pebble skin showing their nature. I have made this Note V as Dr. Seligmann's article in MAN, No. 63, page 3, mentions that these nuclei are not abundant.

NOTE VI.

In a cave at Scarborough out of a total of 122 implements excavated in the first day's work, 43 microliths were included in that total. For description of this cave see the *Observer*, March 5th, 1909.

The details of which total are :—

Chert, one fine falcate, example	...	1
Quartz, cores, of many sizes	...	14
Flakes, useful as scrapers, unworked	...	9
Arrow-heads, a nice example	...	1
Cutters or Knives, one new type	...	22
Scrapers, incurved, "worked"	...	5
Scrapers, ordinary, unworked	...	26
Piercer or boring implement	...	1
Microliths	...	43
		122

QUARTZ.

EXPLANATION OF PLATE I.

1. **Cutting Implement**; upper portion smoothed for forefinger; third streak from the top yellow, signs of work on left edge, white quartz. The bulb at base. **Scarborough Estate.**

2. **Cutting Implement**; upper portion on right smoothed for forefinger, rose-tinted white quartz, bulb at base, our largest quartz example, cutting edge on left. **Maskeliya Estate.**

3. **Cutting Implement**: pale milky transparent quartz or crystal, upper ⅔ left side smoothed for grip, the upper right edge cutting. **Ormidale Estate.**

4. **Cutting Implement**; edge on the upper left side, the whole of the right side flat, white quartz; **from Scarborough Cave.**

5. **Cutting Implement**; rest for forefinger on upper left side, a natural cleavage, the upper ⅔ right side worked sharp for cutting, slightly smoky crystal. **Maskeliya Estate.**

6. **Cutting Implement**; finger rest in centre of right side, natural edge of the flake struck away, pure limpid crystal. **Maskeliya.**

QUARTZ PLATE I.

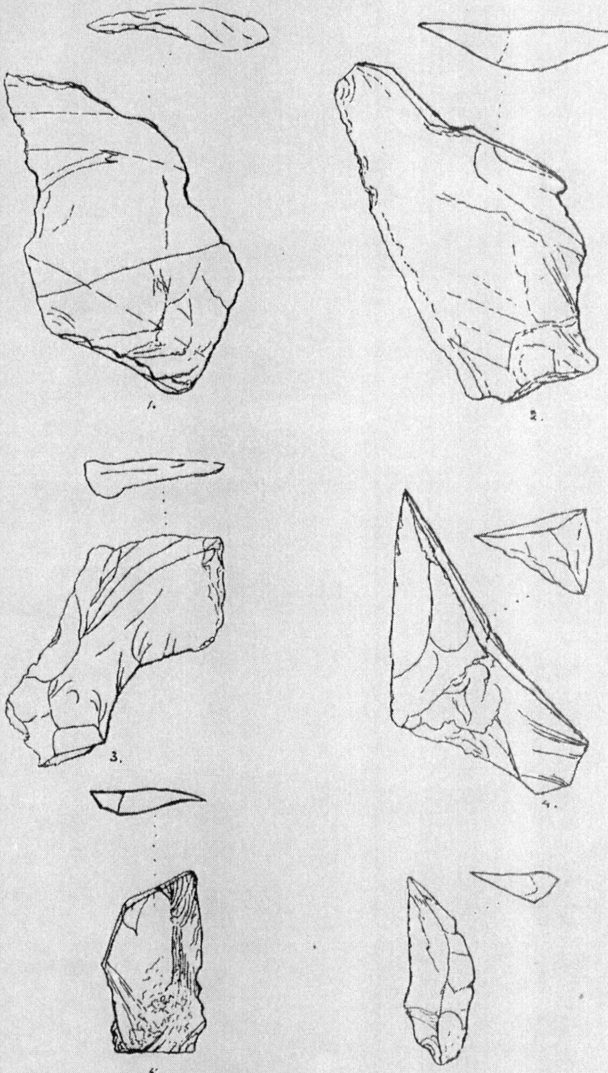

Del. J. P.

QUARTZ.

EXPLANATION OF PLATE II.

1. **Cutting Implement** or Saw, showing the right and left chipping of the palæolithic period, semiopaline quartz, cutting edge on left. **Maskeliya.**

2. **Scraping Implement;** the right edge worked for stave scraping, the lower edge for convex scraping, forming two implements, pure limpid crystal with yellow skin on lower third portion. **Deeside Estate.**

3. **Scraping Implement**; edge of about 80° on the left, black Tourmaline. **Blair Avon.**

N.B.—This appears to have been flaked from its proper mass in the ordinary manner. Found by Preston Plumridge, Esq.

4. **Scraping and Cutting Implement;** concave scraper on lower, cutting by the upper, left edges, pure limpid crystal. **Glenugie Estate.**

5. **Scraping Implement** with **guide on the right base;** from this lobe upwards the whole right edge has been worked at an angle of 45°. The base and left side are both flat, the latter slightly sloping; the upper right edge is sharp, of slightly milky quartz. **Maskeliya.**

There is a large series of this last form of implement, each with guide, as it were to prevent the scraper leaving the "work." Some, indeed, have a double guide, the working surface being countersunk as in a notch, the edge between the notches, generally, very keen.

QUARTZ

PLATE II.

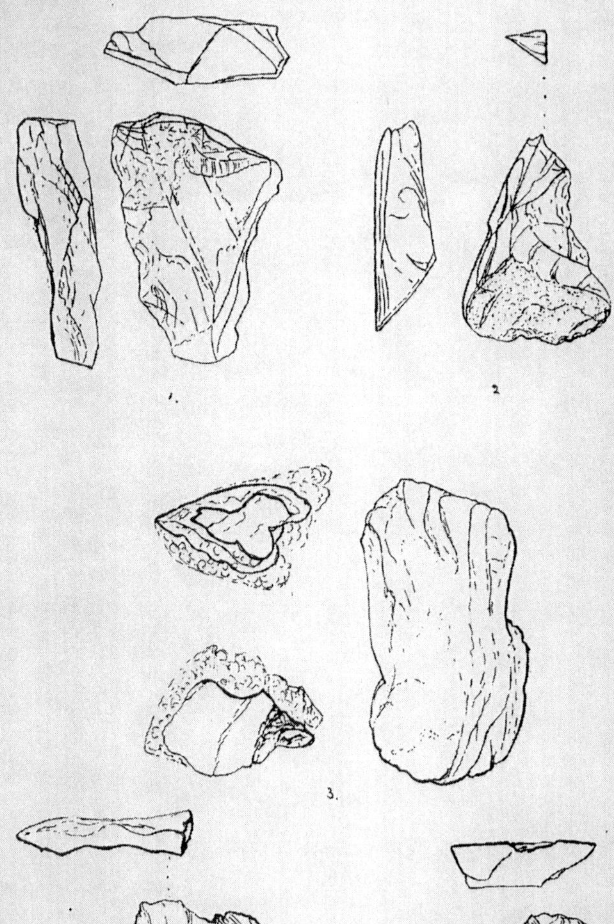

1. 2. 3. 4. 5.

Del. J. P.

QUARTZ.

EXPLANATION OF PLATE III.

Fig 1. **Cutting Implement**; upper right portion nicely "worked" from the apex to the point; entire lower edge keen; pure limpid crystal. **Maskeliya.**

Fig. 2. **Double Pointed Piercer**; probably for sewing hides, etc., whole of the upper portion, from point to point "worked.' Of "frosted" crystal. **Gangawatte Estate.**

Fig. 3. **Cutting Implement**; upper portion, left $\frac{2}{3}$, worked very roughly, lower left half fairly keen; white quartz. **Mankulam Northern Province.**

Fig. 4. Similar to No. 2 except for the single point. **Gangawatte.**

Fig. 5. **Microlith**; the upper side "worked;" the lower edge keen. **From Scarborough Cave.**

Fig. 6. **Boring Implement**; similar in general outline to European examples; white quartz. **Maskeliya.**

Fig. 7. Scraper, incurved, left and right upper sides "worked;" lower edge keen; pure limpid crystal. **Glenugie Estate.**

Fig. 8. **A Saw-like Implement** of semi-opaque crystal; whole of upper side "worked;" lower edge well "worked." **Maskeliya.**

Fig. 9. **Graving Tool**; upper side "worked," lower edge and point keen. **Gouravilla.**

Fig. 10. **Microlith**; upper edge "worked," lower edge and point keen. **Gouravilla.**

Fig. 11. Similar to No. 9, whole of back "worked," lower edge keen. **Glenugie Estate.**

QUARTZ

PLATE III.

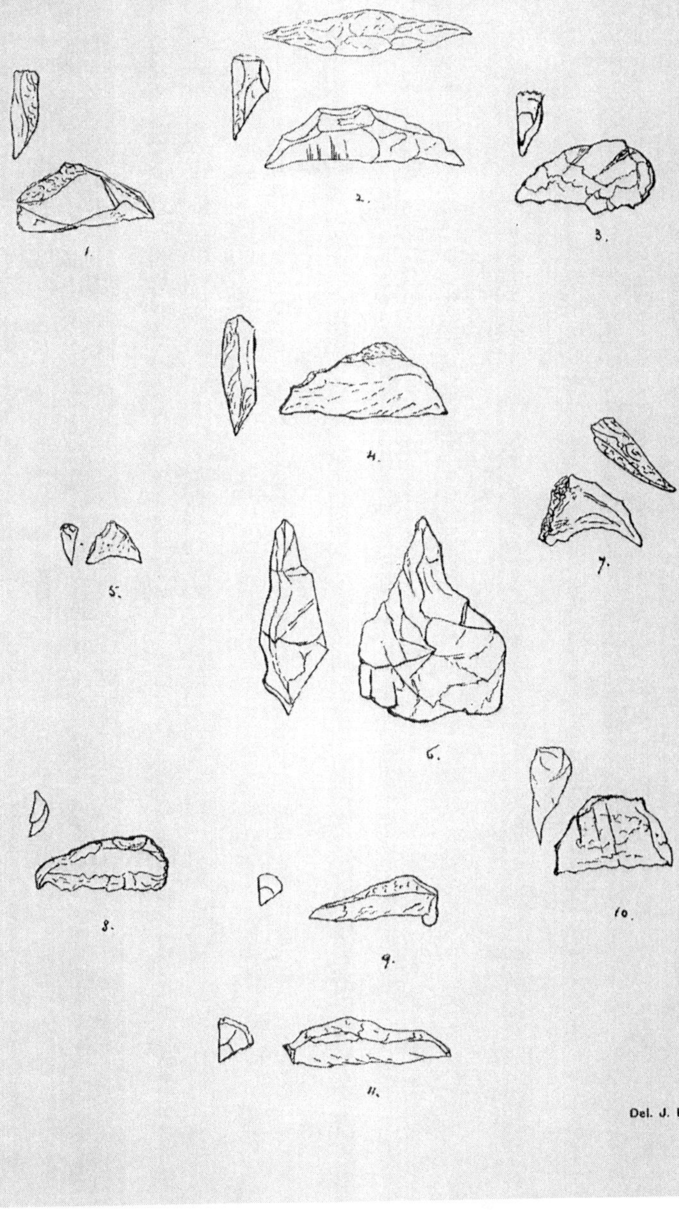

Del. J. P.

QUARTZ.

EXPLANATION OF PLATE IV.

Fig. 1. **Arrowhead** of clear quartz; two edges fairly keen; both "worked;" **Mottingham.** Found by Preston Plumridge, Esq.

Fig. 2. **Arrowhead** of white quartz; the only example with tang as yet found in Maskeliya, both sides roughly worked. **Scarborough.***

Fig. 3. **Arrowhead** of cloudy opaline quartz; edges and point fairly sharp. **Matale, Clodagh Estate.**

Fig. 4. **Arrowhead notched;** middle and upper left portion showing pebble skin; point and edges fairly sharp. Clear crystal Alton. Found by Mrs. C. M. Henry.

Fig. 5. **Spearhead** of clear crystal, much rolled; with a frosted appearance over all but the upper edge. Midlothian. **Presented by C. E. Miller, Esq.**

Fig. 6. **Point** of pure crystal, yellow oxide of iron staining the two lower fasciæ, probably in use as arrowhead, both edges fairly keen. **Brownlow Estate.**

Fig. 7. **Point** of limpid crystal, worked on the left edge; right edge showing natural cleavage, both edges keen. **Uva. Presented by C. Hartley, Esq.**

Fig. 8. **Arrowhead** of white quartz stained with yellow oxide of iron, both sides are worked and fairly sharp. **Maskeliya.**

Fig. 9. **Arrowhead** of pure limpid crystal, very slightly opaline; left edge worked, both edges and point fairly sharp, base notched on both sides, forming almost a tang. **Stockholm Estate.**

---

* N. B.—This is figured by Dr. Seligmann.

QUARTZ

PLATE IV.

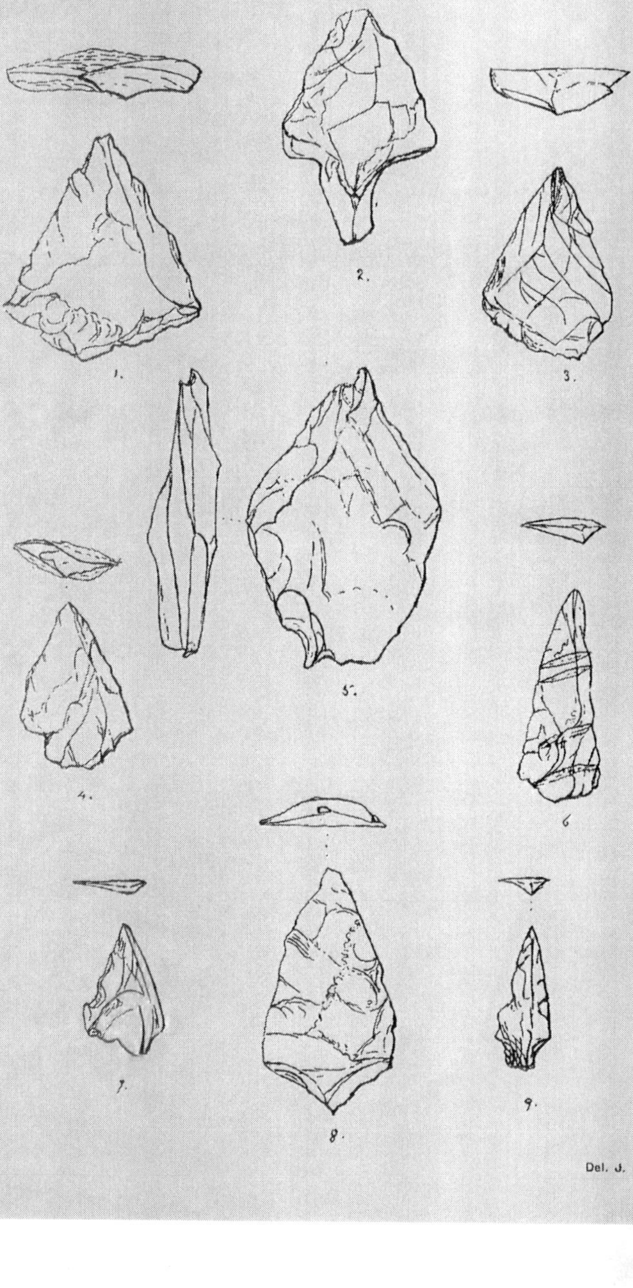

Del. J. P.

QUARTZ.

EXPLANATION OF PLATE V.

Fig. 1. **Scraper or knife** of pellucid crystal, upper edge "worked" for finger rest, lower edge very keen; left side "worked;" may have been used as a scraper with guide. **Deeside. Presented by G. P. Cann, Esq.***

Figs. 2 & 6. (Obverse and reverse) white pellucid crystal; on the left edge are 4 distinct indentations leaving 5 distinct teeth (see also the lower fig. No. 6). This toothed edge is fairly keen. This Dentate implement might have served as a saw. The "work" is more clearly seen on the right edge of No. 6. **Maskeliya.**

Fig. 3. **Dentate Implement**, probably for scraping. **Scarborough.**

Fig. 4. **Cutting Implement**; easily mistaken for Arrow-head. Pebble skin, yellow, on the left $\frac{2}{3}$ right side is the finger rest, quite possibly a broken arrow-head.

Fig. 5. **A pointed flake** from Mankulam N. P., white quartz; edge keen all round except the butt end, the base.

Fig. 7. **Cutting and Scraping Implement;** the upper side is the butt end, all other sides are sharp and bear signs of having been of service, the left edges for cutting with forefinger rest on the top; the lower right basal portion a concave scraper. Pure pellucid crystal from **Forres. Presented by E. H. Etches, Esq.**

Fig. 8. **Cutting and Scraping Implement;** formed from the bulb and hollow of percussion, pure pellucid crystal; right edge fairly flat; left and lower edges keen. **Bargany Estate.**

---

* This is one of a series with Fig. 5, Plate II, but with this difference—Fig. 5, Plate II has a very high angled edge and could never have been used as a cutting implement, whereas Fig. 1, Plate V, has as keen an edge as could have been desired.

QUARTZ

PLATE V.

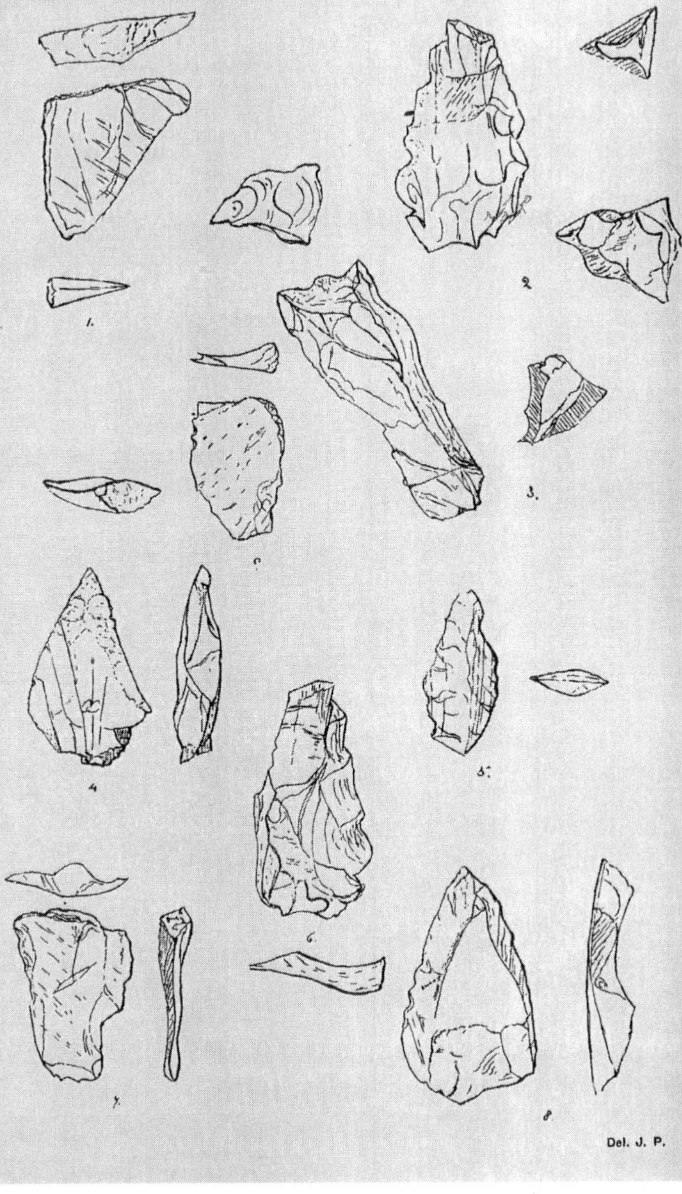

Del. J. P.

BIBLIOGRAPHY.

Old Chipped Stones of India ...	LOGAN.
Palæoliths near Madras... ...	FOOTE.
Prehistoric Annals of Scotland ...	DANIEL WILSON, LL.D.
Ancient Stone Implements of Great Britain	SIR JOHN EVANS, F.R.S., F.S.A.
The Stone Implements of South Africa	J. P. JOHNSON.
Ightham	F. Y. BENNETT.